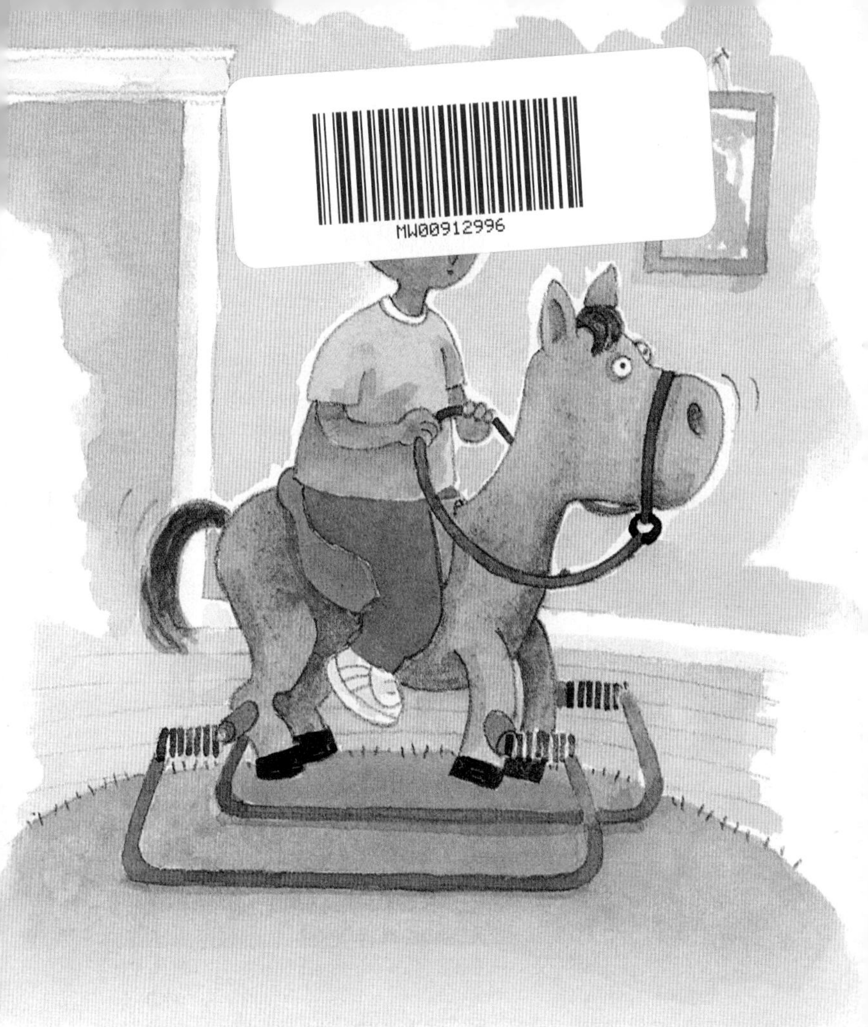

I sit on it.

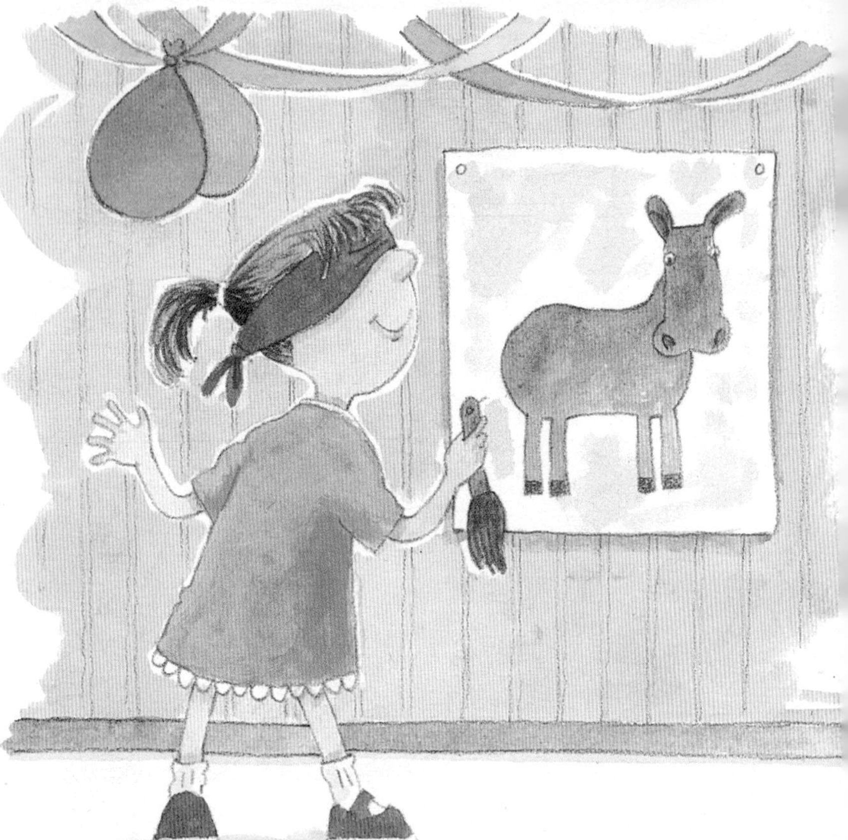

A pin is in it.

Liz pins it on.

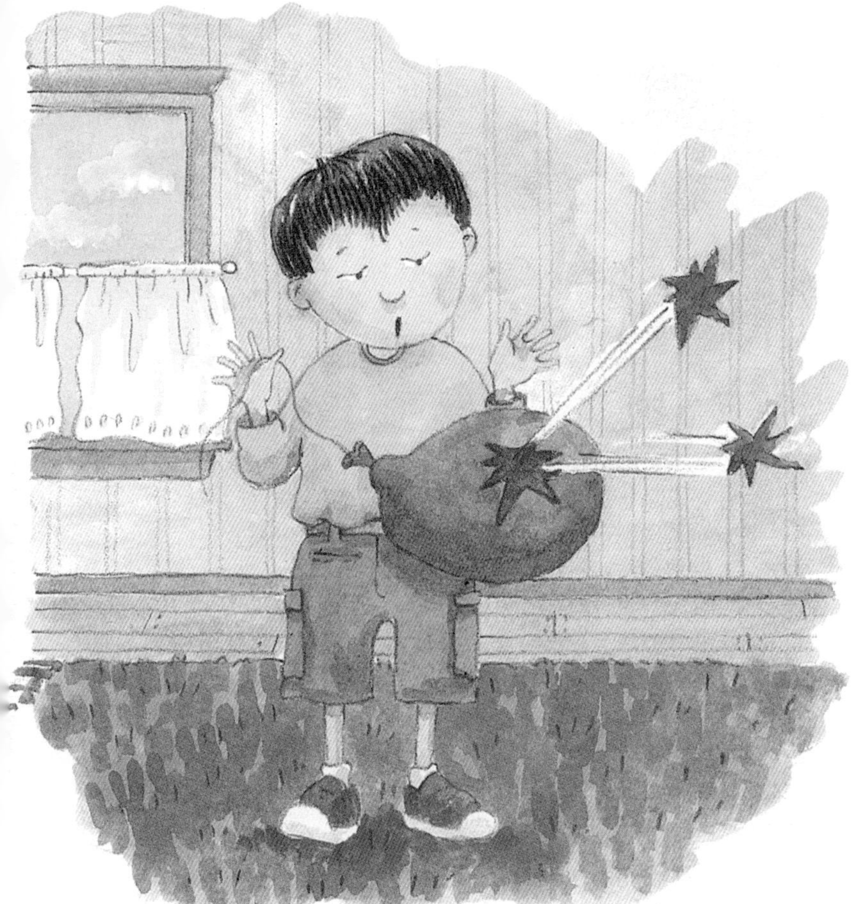

No, no! It pops!

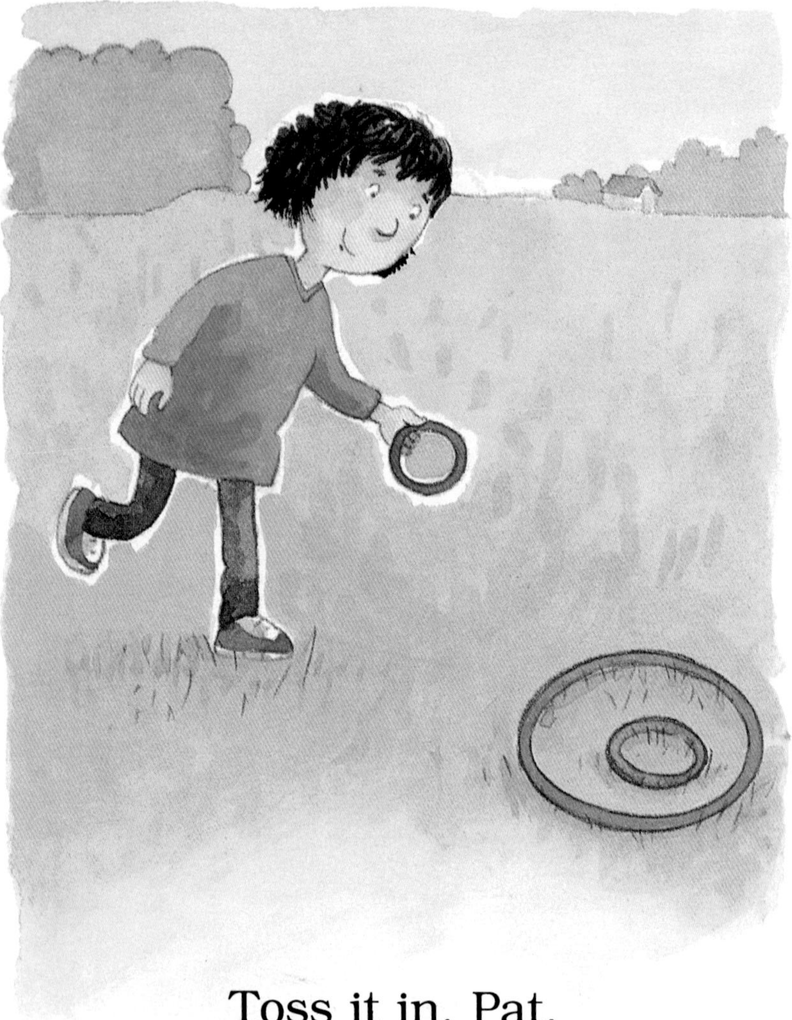

Toss it in, Pat.

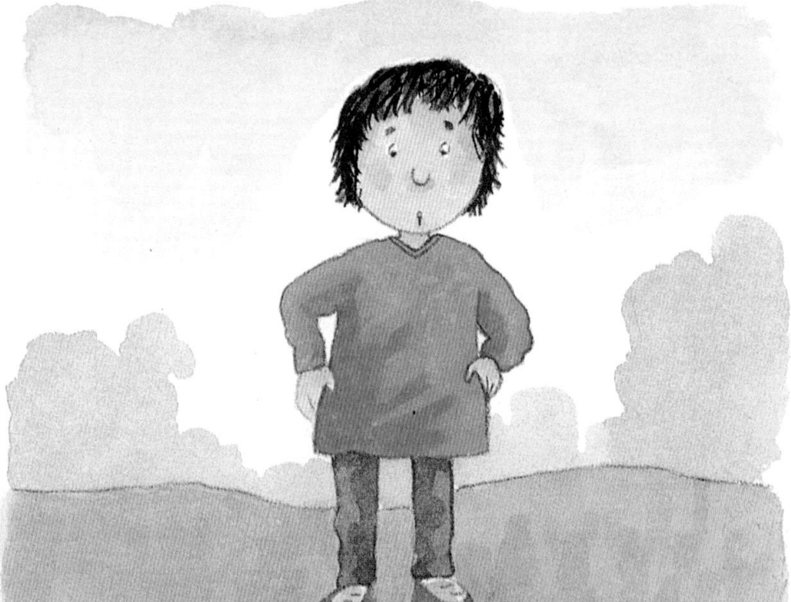

No. It is not in.

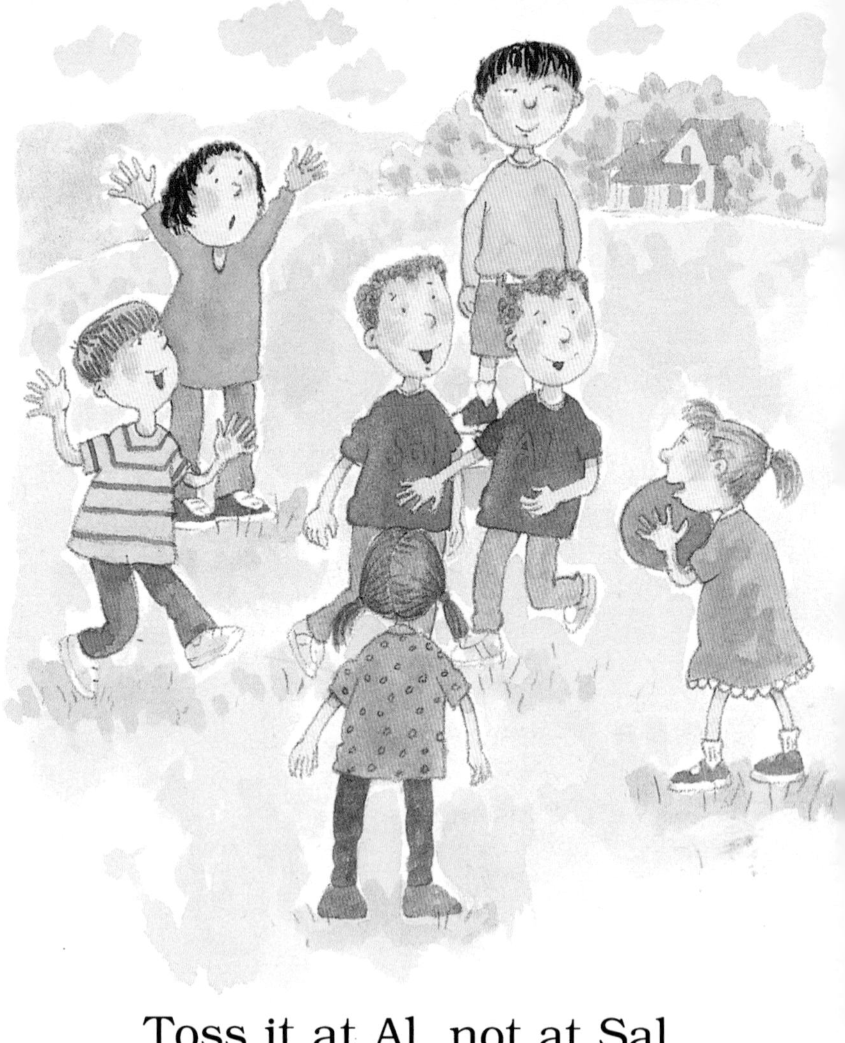

Toss it at Al, not at Sal.

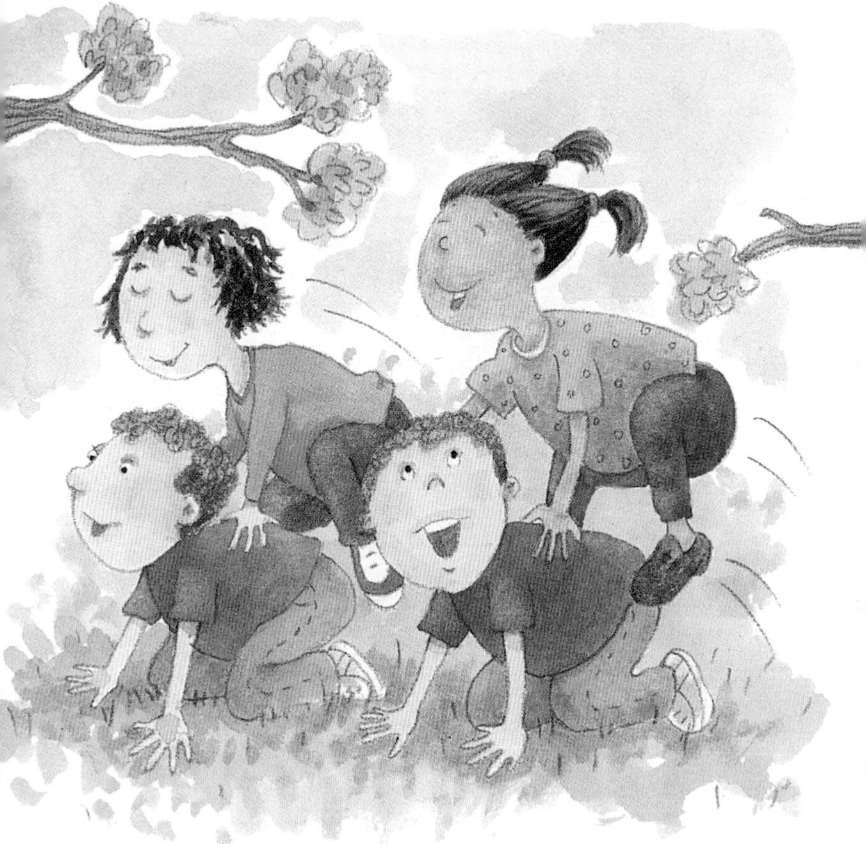

Nan is on top.

So is Pat.

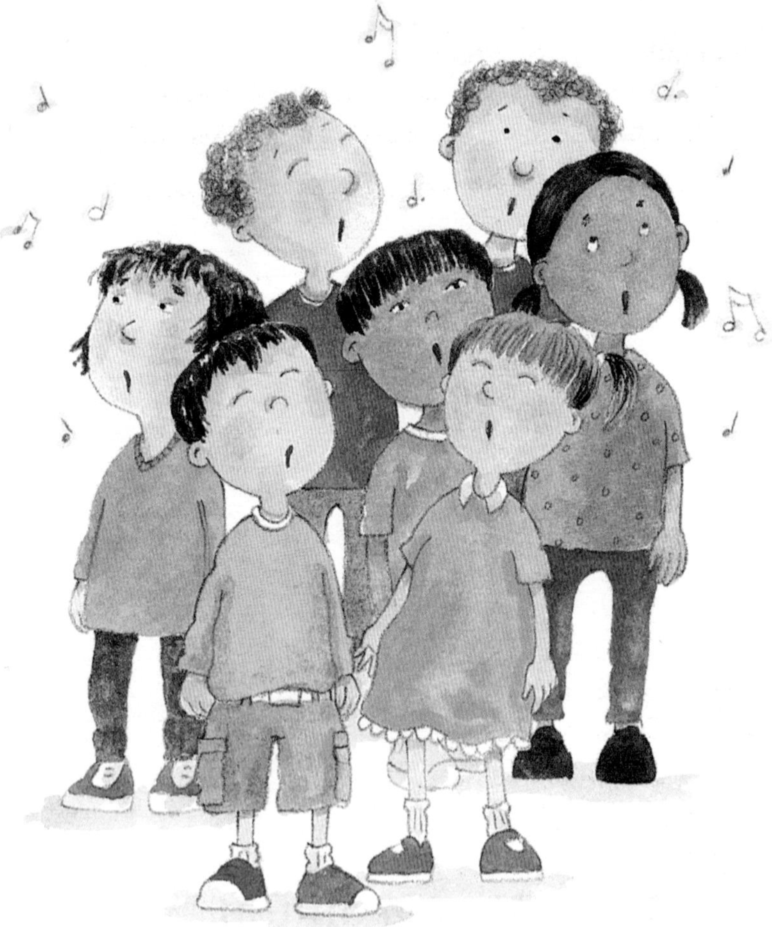

**The End**